# Ham R

*The Ultimate Ham Radio Quick Start Guide – From Beginner To Expert*

**Jack Campbell**

© Copyright 2016 by Jack Campbell

All rights reserved.

In no way is it legal to reproduce, duplicate, or transmit any part of this document in either electronic means or in printed format. Recording of this publication is strictly prohibited and any storage of this document is not allowed unless with written permission from the publisher. All rights reserved.

The information provided herein is stated to be truthful and consistent, in that any liability, in terms of inattention or otherwise, by any usage or abuse of any policies, processes, or directions contained within is the solitary and utter responsibility of the recipient reader. Under no circumstances will any legal responsibility or blame be held against the publisher for any reparation, damages, or monetary loss due to the information herein, either directly or indirectly. Respective authors own all copyrights not held by the publisher.

**Legal Notice:**
This book is copyright protected. This is only for personal use. You cannot amend, distribute, sell, use, quote or paraphrase any part or the content within this book without the consent of the author or copyright owner. Legal action will be pursued if this is breached.

*Jack Campbell*

**Disclaimer Notice:**

Please note the information contained within this document is for educational and entertainment purposes only. Every attempt has been made to provide accurate, up to date and reliable complete information. No warranties of any kind are expressed or implied. Readers acknowledge that the author is not engaging in the rendering of legal, financial, medical or professional advice.

By reading this document, the reader agrees that under no circumstances are we responsible for any losses, direct or indirect, which are incurred as a result of the use of information contained within this document, including, but not limited to, —errors, omissions, or inaccuracies.

*Ham Radio*

*Jack Campbell*

# Contents list

## Page

Introduction..................................................7

Chapter 1: What Is Ham Radio? ..................9

Chapter 2: Key Concepts - Beginner..........21

Chapter 3: Advanced Concepts..................29

Chapter 4: Your First Contact....................37

Chapter 5: Morse Code..............................59

Chapter 6: Assembling Your Own Station......................................................67

Chapter 7: Operating in Bands of HF.......75

Chapter 8: DX - What Is It?......................83

Chapter 9: Failures Common In Amateur Radio.......................................................89

Chapter 10: Operational Technical............93

Chapter 11: Several Expressions Conventional..............................................101

Conclusion................................................107

*Jack Campbell*

# Introduction

I want to thank you and congratulate you for downloading the book, *"Ham Radio - The Ultimate Ham Radio Quick Start Guide – From Beginner to Expert"*.

This book acts as a first-time primer for complete beginners as well as a resource for amateur radio enthusiasts who would like to add to their knowledge base. Contained within this book are some of the most common concepts within ham radio, why they are important, and some applications using real-world ham radio scenarios.

Getting into ham radio doesn't require an engineering degree or months of studying. Once you're able to understand how radio transmission technology works, it's fairly easy to get started with ham radio and begin communicating with other amateur radio operators all over the globe!

There are no prerequisites for reading this book and there won't be a test at the end. It's a straightforward, easy way to get into the hobby.

Thanks again for downloading this book, I hope you enjoy it!

*Ham Radio*

*Jack Campbell*

# Chapter 1

## What Is Ham Radio?

In a remote village of the former Soviet Union, a youngster celebrates their first contact made with a country in South America using your own ham radio equipment, designed and assembled by him. We are in the mid-1980s and the strict standards of the Soviet system to limit the reported information like name, city operator, received signal and local temperature.

Thousands of kilometers away, the experienced South American amateur radio operator launches the contact in your communication record book (QSO) and begins to draw in your mind how will the culture, life and the daily routine of his Soviet mate.

The situation described above seems to be unimaginable in today's world where communication is instant via the internet, mobile phone, satellite and global television networks. But even in modern times, every day millions of amateur radio operators continue to communicate around the world making a large worldwide network of friends.

*Jack Campbell*

By definition, an amateur radio operator is a person who as a hobby, uses a ham radio station for communication without commercial purposes with other people who share the same activity. Depending on the equipment used, such communication may be on your own block or international, or with some amateur astronaut aboard the International Space Station. The communication between hams may be made by voice by digital means, using a computer. Many hams radio operator still prefer to use the oldest means of wireless communication: Morse code or telegraphy.

Ham radio refers to the amateur operation of radio transmitters to communicate with others for a variety of reasons. The term 'ham' in ham radio is derived from what was a pejorative term used to insult amateur radio operators in the 19th century. Insults like 'ham-fisted' or 'ham actor' were directed at and then subsequently adopted by the amateur radio movement, and the term was claimed as their own.

The true Amateur Radio Operator is also a person interested in technical and scientific matters, who really likes to make experiments with antennas, devices, assemblies, etc. Many of the modern communication equipment such as mobile phone and other technological developments are accessible to all thanks to ham radio that allowed these technologies were developed and thoroughly tested.

Most people who get into ham radio are interested in the technology of radio communication for its own sake. The reason for this is because communication via radio over very long distances is a fairly dated method of interacting with people; the internet is vastly superior, after all. So why are so many people operating ham radio setups all over the world?

The popularity of ham radio probably has something to do with its importance in the history of human interaction. The advent of radio communication in the late 18th and early 19th centuries paved the way for a series of massive advancements in technology. There would be huge advantages to whoever was able to master radio communication on the battlefield, too. Entire wars were won because of the ability of one side to be able to use radio communication better than the other, and methods of communicating during an emergency were also enhanced thanks to radio.

The ham radio had throughout its history a very important role in the aid of solidarity situations, disasters and public calamities worldwide. Decades ago when cities were not connected by telephone networks, it was common local amateur radio service support among distant relatives in getting medicines that were only found in big cities or abroad. There are reports that many lives have been saved thanks to the solidarity of amateur radio operator in getting drugs that could not be found locally. Just to cite one example of the importance of this activity as a means of communication in situations of disasters and emergencies, recently, American amateur radio operator, actively participated with an emergency communication network during the terrorist attacks of 11 September.

Think about all of the different ways that you use a radio in your own life. You may routinely listen to the FM radio in your car, and you probably have a favorite radio station that has a specific location on your dial. Also, if you've ever used a cordless home telephone or a remote garage door opener, then you've used some very basic radio communication technology already. In the example of a garage door opener, you push a button that is connected to a battery on the inside of your handheld remote; then power is supplied to a transmitter that broadcasts a specific signal to whatever devices might be listening in range. The motor that is connected to your garage door begins to operate when it receives the signal that was transmitted by your remote.

Because radio waves travel directly through most materials except metal, it's fairly easy to maintain a good read on a radio signal even though you might be inside a building or traveling through a tunnel on the interstate. The methods of communicating using amateur radio all use some format of radio frequency transmission and reception – these two terms simply mean 'sending' and 'receiving.' Some devices can both transmit and receive, and these we refer to as 'transceivers'. When you tune into a station on your FM radio, you're not transmitting any signal. Rather, you are instructing the radio to receive a signal that is being broadcast at a very specific frequency. For example, 99.7 FM refers to the frequency at which all of the music and talking for that station is broadcasted.

But it's not enough to just listen in on what someone else is broadcasting. At some point, most ham radio operators want to actually transmit a signal for others to pick up – they want to be broadcasters themselves. This often requires that they obtain special licenses from the governments under which they are operating so that they can transmit their messages within the confines of the law.

Radio transmissions are often controlled by national governments because they are used for emergency broadcasts, military operations, and other sensitive purposes. Therefore, most governments want to know that anyone who is going to be operating a high-powered radio transmitter above a certain wattage knows what they are doing and what frequencies to avoid using.

Here's an example of what could happen if a radio operator was irresponsible. Say there is a series of traffic intersections that operate on timers that communicate wirelessly to one another. This is a fairly common means of automating stop lights and other traffic signals. If a radio operator were to start broadcasting a signal on the exact frequency that is being used by the timers connected to the lights on the traffic intersections, there could be a potentially dangerous malfunction due to a miscommunication.

*Jack Campbell*

In the United States, it is the Federal Communications Commission, or FCC, that is responsible for managing all of the radio communication that goes on within the country. The radio frequencies that are available for consumers to use vary quite widely and some of them are used for very specific purposes.

The term 'ISM' refers to a series of frequencies that are considered 'license free' – ISM stands for industrial, scientific and medical. Included in the ISM bands are 900 MHz radio frequencies and some 400 MHz radio frequencies. But it's not just the frequencies themselves that are tightly controlled by governments.

There are also limits on the maximum wattage –also known as output power - that can be used in any given transmission. The reason for this is because overpowering another radio network is very possible if someone just decides to add a massive amplifier to their signal, potentially disabling someone else's radio network.

Because all countries have unique laws regarding radio regulation, it's best to do some research first. Check with your local authorities about what radio frequencies are allowable to use, and at what output power ratings. Also, it's probably a good idea to get involved in local ham radio clubs that have members who may know most of this already. They can be great resources for you as you continue to learn more about this great hobby.

## Diversities Hobby

If you are not an amateur radio operator, you have no idea how many unusual and interesting activities you can do. What are the types of people who find? If you walk the streets of your city will certainly find people of all kinds, men and women, persons of different ages, social classes, ethnic groups and religions. They may be engineers, housewives, driver, police, bank clerk. But any of them can be an amateur radio operator who inadvertently you can keep in touch via radio.

The amateur radio is a democratic hobby, which does not tolerate social discrimination, racial or political. It matters little to the amateur radio operator if your mate on the other side does not share the same beliefs or political orientations and much less if it is one or the other race, the ham radio is a huge global community where there are differences and what matters is that all have the same common interest.

As an amateur radio operator, you may have numerous choices of activities and interests. There are amateur radio operators who have a Technician Class license and are dedicated only to speak locally in the VHF and UHF bands in a 65 or 130 miles. Others prefer to operate HF equipment that allows contacts to thousands of miles. Many amateurs like to ride their own antennas, experimenting with new circuits and even set up their own equipment. The operation of low power radio is QRP call fascinates many people who see the limitation of power as a challenge. To talk to other people using the old Morse code is something that still fascinates thousands of amateur radio operators around the world as well as those who like to make contacts in RTTY (radioteletype) or digital modes using one radio connected a computer. The ham radio also allows contacts via satellite (yes, there are exclusive satellites for use by radio amateurs), contacts through the lunar reflection (the signal is folded down on the surface of the moon) or you can even keep in touch with ham radio in the International Space Station (most astronauts is amateur radio operator).

If you have other hobbies as the practice of off-road rallies, navigation or loves to bike trail, you can give your vehicle a modern ham radio equipment that will bring more security and you company on your long adventures.

The ham radio has a code of ethics that was written in 1928 by the American Amateur Radio Operator Paul M. Segal, W9EEA:

1. CONSIDERATE - The Amateur Radio Operator is attentive and never use his station to harm the other activity;

2. LOYAL - The Amateur Radio Operator is loyal and offers their loyalty, encouragement and support to their comrades, to your local club and the American Radio Relay League, through which Amateur Radio in the United States is represented nationally and internationally;

3. PROGRESSIVE - The Amateur Radio Operator is progressive and keeps its station always technologically updated, maintained and properly installed and operating efficiently;

4. FRIENDLY - The Amateur Radio Operator is a friend and patient with other mates, especially if they are beginners. Advises and assists beginners. Provides assistance and collaboration. Considers cooperation with foreign interest. These are the characteristics of the spirit of amateur radio;

5. BALANCED - The Amateur Radio Operator is balanced. Radio is his hobby and he will never allow his hobby interfere with any of their duties and household duties, professional, school or the community in which he lives.

6. PATRIOTIC - His station and skills are always available for service to his country and his community.

## Contests

The competitions are for Amateur Radio Operator, the same as Olympics are for athletes: a stage to show the skill and talent, as well as a stimulus to continually improve our conditions, both operational and technical.

The increase in technique operating and increased operational efficiency are the predominant final results of an operator who always participate in contests, whether a serious competitor whether a possible participant. Any questions about this? Listen bands, any, hear the most efficient operators. You can bet that the vast majority of people taking part always contests.

The contests operator knows from experience to be concise and brief is essential. The contests operator also like to own one of the best signs of the bands. It does not necessarily have the most developed equipment, but a sign that demonstrates efficient use of components available more in your station.

Participate in contests is also one of the easiest ways to contact local and / or stations that need to work to qualify for numerous diplomas. However, one thing that must be understood is that serious operators contests do not like delays, so do not say in the middle of contest that you need the QSL, just cite it in the QSL card that you will send. Nor is it imperative that you be an AZ (in telephony or CW) to participate in contests or even work DX. Just avoid operating near the lower limits of QRG's where usually focus most activities.

Now that you know a little bit about radio technology, how it's used and for what reasons, we're going to dive a little deeper into some of the technical concepts in ham radio. We'll start with some simple definitions of radio terminology, and then we'll look at some of the hardware that is used to operate a very basic radio transmission station. Then, we'll turn up the complexity a bit and introduce some more complicated concepts, before getting into some advanced material more suitable to seasoned radio operators. If you find yourself getting a bit lost while reading this material, don't worry! Take your time and be patient; the rewards for learning ham radio are just around the corner!

*Jack Campbell*

# Chapter 2

# Key Concepts

We're going to start our chapter on beginner concepts by defining a couple of key terms that are considered fundamental to the phenomenon of radio data transmission. We won't just be looking at artificial radio transmissions, either. Did you know that radio frequency (RF) emissions are naturally occurring? More on that later.

**What Is a Radio Wave?**

You might have noticed that I used the acronym 'RF' in the above paragraph. You will see this used multiple times throughout this book as a reference to anything that relates to the transmission of radio information, whether that's an identification beacon, a bit of human speech or simply noise.

RF is defined as any type of electromagnetic wave radiation that lies in the frequency range that extends from 3 kHz to 300 GHz. When we talk about frequency ranges, it's important to note that we are referring to something called the electromagnetic spectrum, which is itself a term used to refer to the total compilation of all potential frequencies of electromagnetic radiation. It's interesting to note that if we climb up high enough in the EM spectrum we arrive at visible light, which is actually a very thin EM section that lies between near infrared and near ultraviolet.

So technically speaking, the visible light that we see with our eyes is composed of the same kind of radiation as the radio waves that are used in radio communication. Because we're only concerned with EM radiation between 3 kHz and 300 GHz, we won't be discussing frequencies outside of this band. Speaking of kHz and GHz...let's go ahead and define what those acronyms mean, too.

## Hertz

Heinrich Rudolf Hertz was the first person to prove the existence of electromagnetic waves, and his last name has therefor been used as the unit of measurement for EM radiation. Hertz is now used to refer to 'one cycle per second', so a light that is flashing at exactly one time per second could be said to be operating at a frequency of 1 Hertz. If a light was flashing only once every two seconds, can you guess what the Hertz rating would be? The answer is .5 (half of one Hertz).

The Hertz measurement isn't strictly confined to EM spectrum phenomenon. It also extends to the fields of vibration measurement and even computers. But for the purposes of this book, when we use the term Hertz, we are strictly referring to the unit of measurement relegated to the EM spectrum. Here are some common prefixes and their associated Hertz ratings:

| Value | SI symbol | Name | Value | SI symbol | Name |
|---|---|---|---|---|---|
| $10^{-1}$ Hz | dHz | decihertz | $10^{1}$ Hz | daHz | decahertz |
| $10^{-2}$ Hz | cHz | centihertz | $10^{2}$ Hz | hHz | hectohertz |
| $10^{-3}$ Hz | mHz | millihertz | $10^{3}$ Hz | **kHz** | **kilohertz** |
| $10^{-6}$ Hz | µHz | microhertz | $10^{6}$ Hz | **MHz** | **megahertz** |
| $10^{-9}$ Hz | nHz | nanohertz | $10^{9}$ Hz | **GHz** | **gigahertz** |
| $10^{-12}$ Hz | pHz | picohertz | $10^{12}$ Hz | **THz** | **terahertz** |
| $10^{-15}$ Hz | fHz | femtohertz | $10^{15}$ Hz | PHz | petahertz |
| $10^{-18}$ Hz | aHz | attohertz | $10^{18}$ Hz | EHz | exahertz |
| $10^{-21}$ Hz | zHz | zeptohertz | $10^{21}$ Hz | ZHz | zettahertz |
| $10^{-24}$ Hz | yHz | yoctohertz | $10^{24}$ Hz | YHz | yottahertz |

Electromagnetic waves are self-propagating iterations of photon pulses. These pulses have detectable frequencies that allow us to identify them according to how quickly they oscillate. This is why when you tune your radio to a specific radio station, you are matching your radios RF receiver to the exact oscillation rate of the radio station that is broadcasting the music you wish to listen to.

On a very basic level, this is how radio communication works. In order to begin actually doing anything with radio hardware, we still need to understand a few more key concepts so that we're not completely lost when it comes time to being building our first amateur radio setup.

## RF in Other Daily Uses

You may have seen the acronym 'RFID' before. It stands for Radio Frequency Identification, and it's now used pervasively as a way to remotely detect data that is stored on an RFID chip. The way this works is that there is either a passive or an active RF chip that is installed onto something that is to be tracked.

A good example of this is RFID toll badges that are used to pay tolls on highways without the driver having to slow down or stop to pay physical cash. In this case, there is an active RFID chip that is connected to a battery that sits in a small box that adheres to the driver's windshield. When this small box comes into range of a transmitter of what is called 'interrogatory radio waves,' it is suddenly detectable and will be able to be scanned for its inherent information.

In the case of an RFID chip that has been surgically implanted into a dog, the chip is passive and doesn't have a battery. Even still, the chip can be read from close range with a strong enough transmitter of interrogatory radio waves.

Did you know that your home microwave that you use to heat up your food is a source of EM radiation? What about the sun? It too emits massive amounts of EM radiation, but only some of it (a very, very small slice actually) is visible to the human eye.

In fact, the planet Earth is constantly being bombarded with EM radiation that hits us from virtually every angle, coming from celestial bodies elsewhere in our galaxy and the universe. So why doesn't this huge sea of potential noise scramble our signals?

The reason for this is something called amplitude. Amplitude is a term that is used to describe the power of a radio wave, or how strong it is. A radio signal that is broadcasted at an amplitude of only 1 watt will be very hard to pick up compared with the identical radio transmission that is broadcasted at 4 watts, or even 400 watts.

The term 'watt' may be familiar to you as a unit of power measurement, and that's no coincidence: in order to transmit any type of EM radiation whatsoever, there absolutely must be a source of power. In the case of the sun, it's nuclear power. In the case of walkie talkies, it's battery power from, say, a 9-volt battery inside each handset.

In order to add amplitude to a radio signal, power can be applied to whatever transmitting antenna is being used to transmit the signal. We'll talk more about antennas, amplitude and signal gain in the next chapter. For now, just be aware of these terms and know that they are important in the grand scheme of radio.

This chapter was meant to give you an overall introduction to radio frequency and electromagnetic radiation. Getting into chapter three, we'll start to explore some more intermediate-level subject material.

**Radio Frequency Spectrum**

The so-called "frequency" is like a telephone line that you're talking about, that is a communication channel and the channel spacing is the difference between the frequency of two channels and this difference is usually 12.5 kHz or 25 kHz. But the range of a radio is directly connected to its power, the geography of use site and the use or not of repeaters and don't the frequency. You see the RF output power. It is called the "power" of radio, that determine (among other external factors) the radio range. It is measured in Watts, and to facilitate the calculations, every watt is equivalent to about one KM range.

The entire frequency of the radio spectrum is divided into four segments. This division is necessary because waves of similar frequencies can have quite similar properties.

1) High-Frequency (HF): Also known as shortwave, these are frequencies under 30 MHz.

2) Very High Frequency (VHF): encompassing the ranged between 30 and 300 MHz.

3) Ultra High Frequency (UHF): the range from 300 MHz to 1 GHz.

4) Microwave: this is the range above 1 GHz.

A Basic Rig

Knowing all these concepts is useful, but on its own it will not grant you entrance into the world of ham radio. For this purpose, you will need at least a basic radio setup.

This basic setup is often referred to as "rig" by ham operators. It consists of a receiver and a transmitter, enabling you to listen to other operators and communicate with them.

Another important part of the rig is an antenna. Antennas are discussed in more detail in the next chapter, but it is a required part of the equipment, enabling you to pick up the signals.

The communication equipment consists of a microphone, headphones and, when Morse code is used, a traditional straight key. In the modern period, a paddle and keyer are also used for Morse code communication, because it is much faster and simpler to use.

In order to communicate digital data, you will also need a computer to be able to interpret the signal. For this purpose, you can disconnect the microphone (and headphones, if necessary) and plug in the required devices.

**Filters**

Some ham operators use filters to reject certain frequencies or ranges of frequencies. There are several types of filters, depending on what you want to achieve.

1) Feedlines are used to stop the unwanted frequencies from going past the antenna from either end.

2) Receiving filters go inside the radio. They usually consist of quartz crystals and allow only for desired signals to reach the receiver.

3) Audio filters usually help with removing the unwanted noise

4) Notch filters are used to clear out a narrow range of frequencies

# Chapter 3

# Advanced Concepts

Now that you know a thing or two about radio transmission theory let's get into some of the more intermediate concepts. In this chapter, we'll look at antenna types, signal gain, signal-to-noise ratios, and licensing for amateur radio operation.

**Antennas**

Generally speaking, there are three main types of radio antennas:

- Omnidirectional (also called 'omni' antennas)

- Directional (also called 'Yagi' antennas)

- Panel (also called flat antennas)

The construction of antennas is done in such a way as to direct the signal of the transmission in a specific pattern. In the case of an omnidirectional antenna, 'omni' means 'all', and 'directional' refers to aiming of the signal. So, an omnidirectional antenna aims its signal in all directions.

*Jack Campbell*

While this may sound like the best kind of antenna, keep in mind that because the signal is being spread out over all directions, it loses its power. Think of it as having a small pool of water versus a stream of water. The stream of water is going to move more volume and have more power.

Directional antennas are reminiscent of the older style TV antennas that were mounted on home rooftops. They typically consist of aluminum parts, and they have sometimes been referred to as 'fishbone' antennas because their antenna elements resemble a fishbone.

Directional antennas funnel the radio transmission into a single, pie-slice-shaped direction. This is important because, in order to get the best reception for your amateur radio setup, you need to know the direction from which your intended signal broadcast is coming from, and you need to orient your antenna accordingly. This can be done by obtaining GPS coordinates and entering the data into a map that will provide azimuthal coordinates back to you.

Panel antennas are probably not going to become part of any ham radio operators kit because they are typically used in the transmission of microwave data at very high speeds. Most amateur radio operators rely on very high-gain, directional antennas with many elements.

## Signal Gain

As mentioned earlier, it is possible to amplify an RF transmission signal in order to make it easier for the intended recipient to get it. This can be done by using high-gain antennas that do not have amplifiers, or it can be done by injecting power into the transmission line itself, by way of what is called an in-line RF amplifier.

Whatever method is used to boost the signal, the term referred to the change in the signal itself is known as decibels, or 'db' for short. You may have heard the word decibel used before to describe the volume of sound, and that's no coincidence. The way that sound volume scales according to amplitude is almost exactly how RF power is scaled as more wattage is added to the transmission hardware.

A decibel level is just a measurement of a signals power ratio. That's really it...but you will find yourself scratching your head a bit as you do more reading on the subject. That's because decibels are also used in other fields like acoustics and engineering, and the application of the decibel may not seem immediately obvious. In the field of RF engineering, a change in power by a factor of 10 results in a change in exactly 10 dB, or 10 decibels.

The decibel was first used by Bell Corporation during the advent of the common telephone. They needed a way to determine exactly how much signal loss was being realized, and they settled on the decibel as a useful tool. Just about every antenna that you come across as you start shopping for your own amateur radio setup is going to have a decibel rating – keep this in mind as you specify your other equipment.

**CAUTION:** Amplifying certain RF frequencies can be illegal in the country in which you live. Be sure to do your research to determine what frequencies are usable in your area, and what their maximum output power ratings are. Also, never operate a radio unit without having an antenna attached to it. This can cause the unit to overheat due to its power not being expended as it should.

**Signal-to-Noise Ratio**

A signal-to-noise ratio is a measurement of an intended radio signal against the ambient RF 'noise' that is all around us. As mentioned earlier, RF transmissions can come from many, many sources – even distant stars in the galaxy! Because there is such a preponderance of RF noise to contend with, we need to make sure that our radio signal is going to be loud enough to be heard among the noise.

Think of being in a loud room with a lot of people, all of them having conversations with each other. Now say you want to talk over the sound of the crowd to a friend that is standing a few feet away. You'd probably have to raise your voice, right? You'd do this to distinguish your sound from the rest of the room so your friend would be able to hear you better. The reason your friend can hear you well is because there is a sufficient signal-to-noise ratio that has been achieved, and communication can now take place.

## Licensing for Amateur Radio Operators

Due to the Communications Act of 1934, all amateur radio is governed by the Federal Communications Commission, or FCC for short. There are also many international agreements that are in place – these agreements directly impact how an amateur radio operator can communicate with radio operators in other countries. In order to operate any amateur radio, one must obtain a license. Generally speaking, there are three classes of licenses in the United States. Frequency availability is granted to those who have obtained classes of a certain grade, and the higher the class, the more frequency options become available.

But getting these licenses doesn't come for free. There's a fair amount of learning required because a satisfactory score on an exam is necessary to qualify for licenses at each level. Even though the FCC regulates who gets what licenses, the actual exams themselves are proctored by volunteer groups consisting of other, like-minded amateur radio enthusiasts. These exams take place under the guidance of organizations known as Volunteer Examiner Coordinators, or VEC's.

Volunteers do the test administration and then report the results to the FCC, who in turn supply the license to the applicant. Most U.S. amateur radio licenses are good for about ten years before they require renewal,

*Jack Campbell*

and the only thing that would stop you from acquiring a license is if you were a representative of a foreign government.

*Structure of Licenses*

Back in December of 1999, the FCC began rolling out a series of changes in how licenses were distributed. In April of 2000, the number of available licenses for amateur radio operators was culled from 6 to only 3. Also, in February of 2007, the FCC stopped requiring that test takers complete the Morse code aptitude section. All of these changes were made by the FCC to make the licensing system more efficient, and to bring it up to date with the digital age that we all live in today. Even though these modern changes have made it a little more difficult to get an initial license, the steps required to progress into the higher licenses are fewer.

The three license classes are:

- The Technician Class- This is an entry-level license available to most any newcomers to amateur radio. The test that needs to be take in order to earn this license contains about 35 questions, mostly pertaining to radio theory and some standard operating practices. The Technician Class license grants the applicant all of the amateur radio frequencies listed above 30 MHz, which allows the user to communicate throughout most of the United States and some other parts of North America. This license also permits some restricted privileges in the use of short wave radio bands – these can be used for a few, select international communication methods.

The General License Class – This class allows for some operational abilities not granted by the Technician Class license. Probably the most appealing aspect of this class is the fact that it now allows the amateur radio operator the chance to communicate with other countries. To obtain a general license, the amateur radio operator must first obtain a Technician Class license, and then they must pass an additional 35 question exam.

The Amateur Extra License – This class grants the ability to operate amateur radio equipment on all frequencies and in all modes and modulations. The final exam for this class is about 50 questions long, and a satisfactory score on it is needed in addition to the prior two licenses.

If you have more questions about how and where to get your own amateur radio license, feel free to visit www.arrl.org.

*Jack Campbell*

# Chapter 4

## Your First Contact

Now that you have a grasp of all the important ham concepts and have (hopefully) successfully completed your licensing process, it is time to start actually hamming it out. If you are interested in amateur radio, you probably already have your first rig all ready to go and are anxious to get in the action.

There are different ham bands transmitting around the clock, so your ears will be your most valuable tool in these initial stages of the process. Tune into different frequencies and listen to what is going on.

You can check a variety of frequencies using the tuning knob on your radio. Your current frequency will be shown on the display or the dial. Depending on the type of signal, tuning method will vary. For FM or Morse code signals, you can use your ears to listen in. However, if the signals are sent using special equipment, you will need a display to get the frequency just right.

10, 12, 15, 17 and 20 meters usually allow great QSOs (reported) long distance (DX) throughout the day. In

30, 40, 80 and 160 meters are also possible great DX contacts during the night, with an emphasis on sunrise and sunset, but during the day, the scope of QSOs is relatively small.

**Common signal types**

Morse code (CW)

Morse code, often also called Continuous Wave (CW), can be tuned into by following this simple procedure. First, set your rig to receive Morse code (switch it to CW mode). Then, if you have more filters, set the rig to utilize a wide filter. Keep adjusting the tuning knob until you hear the signal. Once you find the transmission you are interested in, you can once again apply narrower filter to remove any noise and make for easier listening.

Single-sideband (SSB)

This is probably the most common mode utilized for voice transmission on HF bands. SSB is very efficient and saves radio spectrum space. In order to tune into an SSB signal, first set your rig accordingly. Next, apply the widest available filter. Keep turning the dial until you find a frequency you are interested in.

You can usually figure out you are closing in on an active SSB signal by hearing crackling or rumbling. From that point on, continue fine-tuning until the voice becomes clear.

Frequency modulation (FM)

FM is the most popular transmission mode on the UHF and VHF bands. To tune in, you will first need to set your rig to receive FM signals. Then, you have to reset the squelch in order to ensure you will hear even weak signals, but avoiding the noise.

Now you are set to go hunting. You can enter the frequency using the keypad or search for it using the tuning dial. Once you encounter an active frequency, you will hear the operator's voice. You might have to do some back and forth before getting to the frequency where the voice is the clearest.

Radioteletype (RTTY)

RTTY is a special type of signal. Characters are transmitted as a sequence of two tones: mark and space. These signals cannot be decoded manually (by ear), and they require special external equipment for listening in.

To tune into RTTY, you will first need to set your radio accordingly (RTTY or DATA mode). Next, you must configure the data decoder/encoder. This will depend on the type of equipment you use, and you will probably need to consult the manual to get it right.

When everything is set, you can tune in using the same procedure as for SSB, but instead of listening, you will need to keep an eye on tuning indicator and observe when the tones align properly.

Tuning into an RTTY or DATA signal is a bit more advanced and requires some additional equipment, so you will probably want to wait out on testing this until you have some experience.

## HF, VHF, and UHF

High frequency (HF) bands are organized into two basic segments. The lower segment is usually reserved for CW and data (including RTTY) transmissions while the higher end of the frequency is occupied by voice transmissions. If you are looking or long-distance contacts, you will probably find them in the lower segment. More casual conversations, known as "ragchews," are found at the higher end of the segment.

Contacts on VHF and UHF bands are created with the help of repeaters. The bands are organized into sets of channels. The majority of ham operators transmitting on these frequencies use the FM mode (frequency modulation). It is a great choice for voice signals because it enables noise suppression and thus provides for a good listening experience.

Thanks to repeaters, ham operators can cross large distances with their signals. Often, these communications are more of personal nature then those found on HF, as they are used to stay in touch with family and friends. In order to make use of a repeater, you will need to enable the tone access on your rig. Using this method, you are letting the repeater know that the signal is intended for it. Otherwise, it will not transmit your signal further.

If you want to try and tune in into a repeater in your area, you need to follow these steps:

1) Find a repeater near you. You can do this via a directory or use a simple web search, which should provide you with the list.

2) Find out what are its frequencies for incoming and outgoing signals.

3) Adjust your radio accordingly, setting it up so it listens on the repeater's outgoing frequency.

4) Start tuning as you would for the usual FM signals until you find a live signal.

**Types of contacts**

There are three basic types of contacts (QSOs) you will encounter as an amateur radio operator. There are already mentioned casual conversations ("ragchews"), nets and contesting.

As one would expect, casual conversations make up for most of what you will encounter as you go through different frequencies. Where exactly the term "ragchew" (i.e. chewing the rag) came from remains unclear. However, it has become a generally accepted name used among ham operators.

Once you decide to finally go out in the wild (which is described a bit later), ragchews will serve as a great training ground. They are informal, and you can talk about any topics of your choosing. This will help you build some much-needed confidence and experience.

*Jack Campbell*

Nets (networks) are meetings of ham operators that take place on a consistent basis and at prearranged times. Nets revolve around a certain topic or set of topics, and you can find many of them dealing with rig operation, covering technical and other related questions.

If you have found a specific net you are interested in, you will need to follow certain rules in order to participate. First, check in with the NCS (net control station), which sets the times for meetings, duration, etc. If you want to see what the net is all about as a visitor first, check what their rules on that topic are, at what times they accept visitors, and what is expected of you.

Contesting is a very interesting form of contact. These are different competitions in which ham operators try to exchange short messages and their call signs as quickly as they can. Likewise, chasing DX is another form of a contest in which one pursues contacts with distant operators and stations.

These are some basic rules to follow when participating in a contest.

1) Make your contact brief. The goal is making as many contacts as possible in the shortest time span, so you want to keep the communication short.

2) If DX-ing, exchange the minimum necessary information and move on to the next station.

3) Before getting into a contest, take the time to listen to a particular station and see what their contest is all about, i.e. what type of information they typically exchange. Sometimes it will be your location and the signal report, at other times it can be a serial number, etc.

4) You are communicating with people from all over the world, which means that you are bound to encounter some who do not speak English. They will usually know enough to be able to exchange the basic information, but if they do not, you can use the international set of Q-signals.

## Q signals

Although most people do speak at least some English these days, learning at least some Q-signals can be very useful as your last resort. These are abbreviations originating from the early days of the radio, and every amateur worth his or her salt should be acquainted with them. They are often used between English speakers as well, to make the communication faster.

Here, we bring just a few of them for your reference. A full list of Q-signals can be easily found online with very little effort.

QRG - asking what is your exact frequency?

QRN - are you receiving static?

QRT - should I stop sending?

QRV - are you ready?

QRU - out of things to talk about.

QRZ - who is calling?

QTH - what is your location?

QRX – Will call you again

QSB - Fading of signal

QSK – Break-in

QSN – I heard you

QTC - Traffic

QSL - has my transmission been received and understood?

Making your first call

When we started on the radio, we normally spend much of our time listening QSOs in a progress of other operators. You learn a lot, but nothing like being involved in QSOs to quickly increase our operational efficiency. When you are involved in a QSO, the information you receive is much more important than a simple reception and training, so you will strive to get it right. In the HF bands (High Frequency) there is a high degree of understanding and help. Do not bother to go to the air and start operating. The initial moments of sheer terror quickly turn into panic and this will become a great and pleasant expectation continuity.

Now that you are acquainted with how different types of conversations function and how you can locate them, it is finally time to make your first call. By this point, you should feel relatively comfortable and quite certain you will not be overwhelmed by what you encounter out there.

The station transmits a CQ (general call to all stations), possibly will be several amateur radio operator answering your call, providing chances contacts to the caller. It is more productive to have multiple stations responding to your CQ than you be one of the many stations that respond to a QC. The easiest way to start is to find somebody making a general call, open for everyone. You will recognize this type of contact by the operator calling CQ. It is time to get involved.

Listening

One of the most important conditions that should develop to become good amateur radio operators is the reception.

There is a big difference between operators who expect answers to their CQ's and listen carefully to the answers and those who simply do not hear answers.

At the end of a CQ, carefully check the frequency above and below its own QRG transmission because many answers are transmitted to +/- 2 kHz (kilohertz) of QRG in the CQ that was broadcast.

46

*Jack Campbell*

If no response was heard, repeat the CQ. Keep this sequence of calling and listening to a QSO (contact) be established or until you decide to change frequency or even perhaps to QRT (turn off the equipment).

After sending a CQ is always attentive to the answers. Some do QSY (frequency shift) when you do not get answers to their QCs.

Changes followed by CQs followed by QRG are great to cause unnecessary interference. Short series of CQs on the same QRG always bring responses unless the spread is fully closed.

Accustom yourself to always operate with headphones, keeping the audio level as low as possible. This procedure will bring many dividends in the future, because while you develop your hearing also avoid hearing problems in the future. At first, you may feel some difficulty but soon you get used and later a good reception (Your!) Together with a well-developed hearing ability you can get that call (rare season) distant that no one heard. Only you!

You also need training to tune in a station that answers us when another station that is not in contact with us operates in a very close QRG, causing QRM (interference caused by other (s) station (s).

**Answering**

First, you will need to identify the call sign of the operator. You will be able to extract this from the message they are transmitting on their frequency. For example:

- CQ, CQ, this is Golf November Alpha, Mike.

The capitalized words contain the call sign. The initial letters make up the sign (in this case, GNAM), but instead of spelling the letters alone, words are used for easier identification. This is to avoid any possible confusion between similarly sounding letters like "B", "P", "T", etc.

Once you have identified your CQ, turn on your microphone and send out your message, citing their identification, and naming yours, for example.

- Germany November Alpha May, this is Sierra Romeo India Mike, over.

You can repeat your message a few times and then wait for the response from the operator. They will respond repeating your identification, identifying themselves once again, and citing your signal report.

And that's it; you have just made your first QSO!

**Note**: sometimes you will fail to make contact on your first attempt, or even first few attempts. Don't be discouraged by this. Perhaps your signal is too weak to reach out, or maybe there is a strong interference at the target station. Just keep browsing and you are certain to make some contacts.

However, if you keep failing, it may be time to check your equipment. If you have a licensed ham operator friend, the easiest way to check if everything is working properly is to have them try to make contact. If this is not an option, check following things:

1) Transmitting frequency: to check if you are sending out the signal on the right frequency, press the "speak" button on your microphone (or press the Morse key) and observe the radio display. If the indicators for the sideband and frequency do not stay the same, you are probably transmitting on a different frequency.

2) Check your transmitter's power, by observing power output meter.

3) Make sure your antenna is connected correctly. If there are no incoming signals, or if they are very weak, there might be a problem with your antenna or the cable.

## International Phonetic Alphabet

| A | Alpha | N | Novmber |
|---|---|---|---|
| B | Bravo | O | Oscar |
| C | Charlie | P | Papa |
| D | Delta | Q | Quebec |
| E | Echo | R | Romeo |
| F | Fox Trot | S | Sierra |
| G | Golf | T | Tango |
| H | Hotel | U | Uniform |
| I | India | V | Victor |
| J | Juliet | X | XRay |
| K | Kilo | W | Whiskey |
| L | Lima | Y | Yanke |
| M | Mike | Z | Zulu |

## What information to exchange?

Now that you've finally made a contact, you might be in a bit of a dilemma as to what to talk about. This is not unusual since you are talking to someone you don't know and starting and maintaining a conversation may be difficult, especially if you are a bit more shy and not as talkative in general.

With time, you will get better at it. Like everything else, it takes some practice and getting used to, but for new ham operators, here are some tips.

You can exchange information about the signal strength, give your name and location. This way, you will also find out some basic info about the person on the other side and will be able to determine if the circumstances warrant for a longer conversation.

The information about your rig is also a good conversation starter. You can talk about the type of rig you are using, what antenna(s) you have, etc. And, there is the good, old topic of weather - starting conversations since the humanity learned to speak.

If you get a good feeling about the person on the other side and feel like he or she is someone you can communicate with more, you can exchange other information, like your age, hobbies, job, etc. Basically, anything you would talk about with any other acquaintance. Remember that ragchews don't observe any particular set of rules pertaining to allowed topics, so just talk about anything you feel comfortable with.

Three subjects that you may want to avoid speaking about on air are politics, sex and religion. These are, generally speaking, sensitive topics, so you should probably steer clear of them and avoid getting into these conversations when you happen to encounter them while tuning.

## Retention Frequency (QRG) of Transmission

It is natural we move the main line of our transceiver when we are listening for an answer to our QC. This is the correct procedure when we have frequency adjustment control for RX (receive) and TX (transmission) apart. However many new amateur radio operators operate transceivers so that tuning adjustments affect both the reception when the transmission.

If you use a transceiver, do not move the main line after sending a CQ. If you do this, the other operator will not hear your responses due to its new QRG transmission when you are responding to its defense.

If the change is not big you probably will be heard and will be able to proceed with the QSO but one of you will be in a different QRG of the other which is not very recommendable.

## Reception Line

Many transceivers provide conditions to modify the QRG's reception, just it, several KHz (kilohertz) above and below the QRG transmission without affecting this.

This device is known by various names such as BFO (Beat Frequency Oscillator), Clarifier, OT (off-set tuning), Pitch, RIT (Receiver Incremental Tuning) and Tone. No matter how the name on your device but you know the correct way to use it. When tuning in a free QRG, calling CQ or tuning into a station, this adjustment should be placed in the Zero position (or off).

If this is not done, you will not hear the same QRG that will pass. After sending a CQ, do not touch the main line. Use the clarifier to tune in stations that are perhaps answering your CQ. Obviously this is not considered when operating a receiver and transmitter with controls for TX and RX separately.

If you operate a transceiver that has not clarifier (or a separate tuning control) you will notice that it is better to answer to CQ's than send them. This is because the transmission of QRG has to be the same from the other station for you to hear it well and your response should be heard with no or almost no difference in these conditions and if the other operator answering a little out of your QRG, not there way you tune it without changing its QRG transmission.

52

At the end of a QSO, tune your QRG's reception up and down the QRG's transmission that you used and to listen to stations that may be calling you.

## Timing your contacts

If you are looking to find some stations for casual conversations, timing your airtime correctly is the key. Weekdays during daylight hours are usually good periods for ragchewing. Do remember, though, if you are making long-distance calls, your daytime is not everyone else's daytime.

Weekends are also good times around the clock, especially Saturdays, but these are also periods when most contests take place. This means that many hams will be preoccupied with other things and may not be up for some good, old ragchewing. On the other hand, a great number of operators on air increase the chances of finding someone willing to talk.

The best signals that indicate another station is in a mood for some casual chatting include a relaxed tone and slow speaking tempo. Find such a station, which also has a good signal, and get involved in a conversation. Alternatively, you can also break-in in the already existing conversation, but make sure to do it properly, without interrupting.

## Confirming The First QSO

When you work a first QSO (contact), let the other amateur operator knows that QSO and QSL card (written confirmation of QSO) are important to you.

In addition to sending written confirmation of the QSO, you can also add other details of it, thanks to your suggestions and QSL. Keep in mind that the amateur to who you send your QSL can use it to request a diploma and are not pleasant to receive our QSL return for not being properly filled.

Try sending a card strikethrough for DXCC! DXCC is the most famous Diploma in Ham Radio, being necessary for obtaining confirm QSOs with at least 100 different countries.

Always remember that the following data are essential in the QSL card so that it is by the book.

1. DATE: Enter day, month and year of completion of the QSO.

2. INDICATIVE: Place the call sign of the mate with whom you kept the QSO, not forgetting to specify any special conditions the same as: mobile (land, sea or air). Some colleagues operating QRP (low power) and like the cards addressed to them mark this condition.

3. TIME: Always work with the international UTC (Universal Time Coordinated) which is GMT + 3 hours. The QTR (time) should be written with four (4) digits, from 0000 to 2359, with no gaps between them, followed or not indicative letter of the zone to which they refer.

4. TRANSMISSION MODE: Put this item the type of emission used QSO, for example SSB or J3A, CW or A1A, RTTY, AM (A3A), etc ..., and some diplomas

54

*Jack Campbell*

require the emission mode is indicated in QSLs as follows: "2 way" or "2 x CW indicates that both involved in the QSO were in CW don't being accepted QSLs otherwise and currently almost all amateurs use of these two expressions.

The minimum reports to internationally accepted diplomas are 338 to CW and 33 to telephony.

5. TRACK OR FREQUENCY: Enter the band (wavelength) or frequency (in MHz) used in the QSO as follow: 160m (1.8 Mhz); 80m (3.5 MHz); 40m (7 MHz); 30m (10 MHz); 20m (14 Mhz); 17m (18 MHz); 15m (21 MHz); 12m (24 MHz); 10m (28 MHz); 6m (50 to 54 MHz); 2m (144 to 148 MHz), etc ...

6. REPORTS: This is a given absolutely essential, as there is no report QSO first during the QSO and then confirmed in the QSL card. In amateur bands is used RST (readibility, strength and tone) and the tone exists only in CW. R varies from 1 to 5, 1 for a signal report unintelligible and 5 to perfectly intelligible. The intensity S ranges from 1 to 9, the same as the S-meter of your equipment. If the signal exceeds 9 you can place a 9 + 10, +15 9 or 9 + ..., etc. In QSO's CW RST report is a 3 digit number which R and S have the same meaning as QSO's in phone number and the third means tone signal CW, T that ranges from 1 to 9, being 1 equal to a CW note modulated by a low frequency signal equal to 9 and a pure continuous current rating without modulation characteristics.

## TIPS:

a) Always sign your QSL cards.

b) Strikethrough Cards give bad impression and are not accepted for diplomas.

c) It is required for most international diplomas a minimum entry, usually 33 to 338 for telephony and CW.

d) Your QSL should clearly state your call sign, your name and full address.

e) Other details such as antenna, power, equipment, CQ zone and ITU zone, grid locator, geographical coordinates can also be added.

If all wait get a QSL card before sending your, would not exist QSL traffic. If you want to get a QSL an accomplished QSO, send your fastest QSL after the QSO.

## Calling On a QRG (Frequency) Busy

It is not correct to transmit a QC at a frequency that is occupied (QRL). This usually happens when someone hears only the last part of a call sign when you are tuning the band. When the callsign is not known with certainty, the operator can send a QC waiting for the another station responds. However, if the amateur radio operator is involved in a contact (QSO) and not calling CQ, your call is likely to cause unnecessary interference.

The station that initially uses the QRG (frequency) also retains the property to continue using it when the end of the QSO.

When you reply to a QC and later complete the QSO remember that he / she has the right to remain using that QRG after completing the QSO.

If another amateur call you when the end of the QSO, the right is to ask it to change (QSY) to another QRG that is not being used.

Usually ask it that he update (frequency) or get off the QRG in use ie do QSY.

**How to request QSY?**

PSE CALL UP or just UP. PSE CALL UP 5, for example, means please call me 5 kHz (kilohertz) above the current QRG.

There are also DOWN, meaning QSY low (lower frequency) but the most used is UP, to be shorter. It is also used UP CALL is a request that the colleague call in the first window (QRG) to free up the QRG in use.

You do not use a frequency that is being used (QRL).

Courtesy generates courtesy and this has been a long time a major point in favor of Ham Radio and everyone expects you to continue with this essential tradition.

## The QRZ USE

If you passed CQ (Caller Line general to all stations) and someone answered but you lost or unable to copy the call sign of station that challenged you, the right way to ask who is calling you is: of PP2 ... QRZ ?

There is not any station in the world whose indicative of Call Line starts with the letter Q. To transmit QRZ? PP2CW (for example) is not correct.

## The QRL Use

A similar incorrect procedure to the example cited above which is also used as a subterfuge to call general is the use of the code "Q" is used to interrogate the QRG (frequency) is busy (QRL) or not.

It is usually used to interrogate the QRG is busy (in use) and get for an answer.

It is advisable to make sure that the QRG is not in use before transmitting. However, a correct example of how this transmission should be called general (CQ) or start a QSO (contact) that the frequency (QRG) is not being used.

Do not use the Q code as synonymous with CQ.

## Transmission Speed CW (Continuous Wave)

Most operators have typically provided forward faster than can receive. This is normal but can cause problems for beginners.

*Jack Campbell*

If you passed up speed that is capable of receiving, this will bring you problems if another operator to challenge (answer) your call forward at the same speed you passed. An effective way to reduce the transmission speed is to focus on transmitting perfect characters. Another way is to increase slightly up the space between words and letters.

It is natural for those who get forward faster than we receive because we are usually trying to increase our speed reception.

# Chapter 5

# Morse Code

Once you have had your fun with voice transmissions, you may become interested in stations transmitting the Morse code (CW). We have already explained on what frequencies you can find such communications and how to tune in. Now it is time to learn what to do once you are in.

First and foremost, if you want to listen to the Morse code signals, you will want to use the headphones, preferably a type that cancels the surrounding noise. This will make it easier for you to zone in. Make sure to set the pitch of the tones to the level that doesn't hurt your ear.

You will probably want to use a filter with a 500 KHz spectrum. Voice filters are usually 2.4 KHz wide, but you do not need this bandwidth for Morse code transmissions. A narrow filter will help cancel close signals and remove the interfering noise.

*Jack Campbell*

### CÓDIGO MORSE

| A ·− | J ·−−− | S ··· | 2 ··−−− |
|---|---|---|---|
| B −··· | K −·− | T − | 3 ···−− |
| C −·−· | L ·−·· | U ··− | 4 ····− |
| D −·· | M −− | V ···− | 5 ····· |
| E · | N −· | W ·−− | 6 −···· |
| F ··−· | O −−− | X −··− | 7 −−··· |
| G −−· | P ·−−· | Y −·−− | 8 −−−·· |
| H ···· | Q −−·− | Z −−·· | 9 −−−−· |
| I ·· | R ·−· | 1 ·−−−− | 0 −−−−− |

## Learning to handle the code

Getting comfortable with CW will take some time and practice, but don't be discouraged. If you set your mind to it, you can do it.

You should first start by just listening to the CW, without writing anything. Since you do not have to respond, there is no pressure. Get comfortable with it, try to hold as many characters as possible in your head. Do not stress during this learning phase. The Morse code is not something that should come naturally to you for any reason, and in many aspects, it is similar to learning how to communicate in a brand new language. Listening and trying to understand is the best starting point.

When you feel ready for it, get a piece of paper out and try to write down the most important information from the transmission. Don't pressure yourself by trying to write down every single letter. It is still too early for this; you are still learning.

As you continue to do this, you will notice that you are making fewer mistakes with every new listening. What is happening is you are developing a skill known as "copying behind," as your brain correct errors and makes sure what you write down makes sense. Keep practicing, and soon you will become fluent at Morse.

## Sending messages out

Since communication is a two-way process, at some point you will want to send the messages out as well. We have already mentioned the standard devices you can use to do this. The straight key is better fitted for more experienced operators, as it requires some effort to make the signal sound good. It can dish out between 20 - 30 words per minute.

Electronic keyers are much better choice these days. They are simple to use, can dish out around 60 words per minute, and make a much clearer sound. If you are just starting out, you will probably want to use a keyer. Once you get more comfortable, and you want to experience what it was like back in the day, you can take a straight key out for a spin.

Nowadays, you can even use computers to extract characters from the code. By attaching an interface to one of the computer's ports, you can sometimes even listen to several transmissions at once. However, computers may experience difficulties transcribing the code if some noise is present.

## Finding the calls

You will find the CW calls on lower frequencies of the HF bands. The lowest frequencies are usually occupied by the fastest operators while the speed decreases as the frequency increases.

First and foremost, find an operator who transmits at a speed that you are comfortable with. You will usually find slower operators in the subbands 3,675 - 3,725, 7,100 - 7,150, and 21,100 - 21,200.

Once you have found the operator you want to make a contact with, it is not all that different from making a contact on a voice frequency. What is mostly different in CW communication is heavy use of abbreviations, for obvious reasons. You can find all these abbreviations online, and you should probably get informed about them before jumping on air.

As you communicate, remember that you are required by FCC rules to include your call sign in a conversation every ten minutes. There are some tips and tricks that you will learn as you go along, turning you into a real CW expert. For example, make sure to always end your contact with prosign "SK," indicating the end of the conversation, or "CL," to message you are going off the air.

**Signs Of Termination**

There has always been much discussion about the correct use of telegraphic signals, including termination signals. All work signs have to mean, and it is bad practice to use others that are not internationally recognized.

The International Advisory Committee for Telegraphy and Telephony (CCITT) and the International Radio Consultative Committee (CCIR) working under the auspices of the International Telecommunication Union (ITU). Decisions of the CCITT and CCIR are published by the ITU, and we usually receive this information through the International Amateur Radio Union (IARU - International Amateur Radio Union).

The difference between symbols indicators such as SK (commercial) and VA (military) has no real meaning since both mean the same thing (end of work) and sound in the same way. SK means you finished a series of transmissions. When you send SK, you are telling the other party that you hear their final comments but does not intend to pursue the QSO.

When the other operating your final comment and also point out that goes away, it is common courtesy to send to GM, GA or NG (good morning, good afternoon or good evening) and compliance and also noted that the final comment was received.

The message end symbol is AR and has no other meaning. If no message was sent or established any QSO, it is not right to send RA, because there is no end to a message that did not exist. The letter K (invitation to transmit) sent out of context (not part of the call sign) and at the end of the transmission tells another operator: REPLY. AR means not respond despite many unofficial publications to indicate with this meaning.

The letter N after the letter K is also inappropriate and unnecessary. Has no meaning accepted internationally, despite being commonly used to indicate that only the calling station response will be accepted. Since that the call sign identification specifies the station you are calling, there isn't need to tell other amateur radio operators do not respond.

It's funny to hear when some stations use KN at the end of a CQ call, for general calls to all stations (CQ) and the end of it is placed a restriction answers: KN (only answer you). Spins are common in telephony but are rare in CW. When a third amateur is trying to put the brakes on a QSO in progress, there is usually a valid reason for termination and I doubt that those who send KN refuse to include a rare DX station or a personal friend to your QSO.

Finally, KN is not a sign of official work.

## UTC - Universal Time Coordinated

It is the old GMT (Greenwich Meridian Time) which is internationally used in radio communications. For each 15 longitudes (west to east) add + 1 hour and - (minus) 1 hour for each 15 longitudes in the opposite direction (east to west).

Converting GMT (PY) to UTC, add + 3 hours. 0000 and 2400 are interchangeable, 2400 is associated with the day's date and ending 0000 the day ahead.

The advantage of the QTR expressed in UTC is that it is used worldwide and it is recommended that the

QSLs, Contests reports (LOGs), diplomas requests are used UTC.

*Jack Campbell*

# Chapter 6

# Assembling Your Own Station

In a life of every ham operator, there comes a point where they want to set up their very own station. It is an amateur radio operator's graduation of a sort. It is the most complicated and the most demanding task that a ham can face. Once you have successfully built your own station, even if it is a simple one, you earn the right to call yourself a real ham. It brings about a special sense of achievement and satisfaction.

1. **Setting goals** - The very first step in this process is figuring out the purpose of your station. There are so many things you can do with a ham station, so the choice is entirely up to you. You should answer some basic questions in your head and figure out what exactly you want to achieve. Primarily, what was your reason for getting involved with ham radio? What attracted you to it? Is there an operator who you admire in the ham community? Do you appreciate what they do, and why? What type of communication are you most interested in (voice, CW, data)? Where will you operate from, will your station be fixed or portable? These considerations will make your future choices much easier. Then you need to consider how

*Jack Campbell*

big your budget is and how much space you have available for your shack and antennas.

2. After establishing the purpose you need to get an amateur radio license. Possession of your license, you will have several options of equipment and antennas, from the simplest to operate on only one track, such as VHF radios to the most sophisticated, such as running band equipment.

> Another interesting option is to set up your own assembly equipment. There is much literature on equipment design in books and on the internet.
> The basic equipment to start a season are:
> - The radio
> - A power supply of 13.8V with current exceeding the maximum consumption of your radio (usually 25amp).
> - A specific antenna for your radio operating range.

3. Determine a free location at the home to the installation of antenna. If you live in a condominium, which guarantees the right to Ham Radio antennas installation.
4. Provide a notebook to record their communications. The reported record book is required in Amateur Radio stations. If you want, there are several computer programs that manage your contacts.
5. Provide a QSL card. The QSL card is a courtesy sent by amateur radio operators and aims to authenticate the performance of your contacts. The card will be helpful in getting diplomas, which require proof of contacts.

The card can be created by own Ham Radio and printed on a computer printer or a graphic. If desired, several clubs offer Amateur Radio QSL cards at a very low cost. They are standardized cards where there is a place to include your code.

And now the fun begins!

## Shopping for equipment

This book will not go into specifics of different types of equipment out there; the list is simply way too big to be included. You can check out what is being offered on the market by going through different ham magazines and browsing online. If there is a radio store nearby, all the better. People working there will be able to help you once you give them the details extracted from the earlier questions.

If you are planning on operating from home, you will not need a lot of space for the equipment. If you are interested in voice modes, just make sure to find a place where your activities will be the least nuisance to other house members (and vice versa - you don't want unwanted noise on your signal, either).

The best places for setting up your rig are usually spare bedrooms or basements, because of minimum temperature fluctuations and dryness. Home operators also require a quality pair of headphones, just in case, so make sure to put that on your list.

Modern technologies have made it possible to have mobile radios that you can carry around easily. You can have them on you at all times, and they make it simple to access all frequencies (HF, VHF, and UHF). However, for HF operators, having an efficient antenna on this type of a device can be a bit trickier.

Finally, you can set up a portable station, and there are numerous ways to approach the construction of this type of a rig. You need to be able to pack your station at all times (including your power source) and move it to a different place. This one is the hardest to plan.

First, you will need to figure out your weight budget. You will have to be creative figuring out your antenna options as well as other accessories. You can get some really small radios that are simple enough to transport and require very little space. The problem, however, is that these radios are often complicated to operate, so if this is your first rig, you will probably want to go for something bigger and simpler. As you become more experienced, you will get to know what features are not absolutely necessary and can make your picks accordingly.

## Choosing a radio

Now that you have all the necessary information, you can finally make a choice for your radio. This will be the most expensive piece of equipment, but don't worry about spending money on this item, as it is the heart of your new station.

If you are buying an HF radio, there are three types to choose from.

1) Basic radio comes with simple controls, a basic receiver filter, and signal adjustments. Displays and metering are quite limited, and there is not much support for external accessories. This is a good pick for a novice. Once you get more experienced, you can use it as a portable radio.

2) Journeyman radio has all the adjustments you could require. Display and metering functions are expanded, and particular models also support digital data and additional bands. There is a computer interface as well.

3) High-performance radios are intended for experienced operators. These radios have a top-notch receiver, full metering and displays, and very extensive interface for computer control and digital data.

If you want to transmit on VHF/UHF frequencies, you will need a radio capable of these operations. There are multi-mode radios, which operate in all modes (SSB, CW, and FM), distinguishing them from FM-only radios.

Practically all hams on VHF/UHF use FM. There are two basic styles of FM radios: hand held and mobile.

Hand held radios can be a single-band or multi-band. While multi-band models cover the range from 50 to 1,296 MHz, they are much more expensive than the single-band types, and you don't really need all those extra bands. The single-band model will be more than enough for most people.

Mobile radios often offer better performance in terms of receivers. They can reject strong signals incoming on the nearby frequencies and are very useful for digital data operations.

In addition to the radio, you will also need a filter to avoid interferences from the nearby signals. For a long time, the only type of filter capable of doing this has been a crystal filter, but lately, there are also mechanical filters, which use vibrating discs to achieve the same goal. HF radios are usually shipped with at least one SBB filter, but you will probably need to get another narrow SBB filter or a 500 Hz CW filter for Morse code communication.

Another piece of equipment you may look into is an amplifier. It will make your signal much stronger, but it is probably for the best for less experienced (HF) users to refrain from buying an amplifier until they garner some experience. The reason for this is that your signal could interfere with other hams, and this is not how you want to start your station.

For VHF/UHF operators, amplifiers can really give you that extra power you need to increase, for example, the output of your hand-held radio. However, you must be weary of safety issues. If the antenna is close to people, you should not use the amplifier at or above 50 MHz, as it can be a grave danger.

Finally, you will need to make your choices regarding the antenna, connectors, and any additional accessories, like a computer, microphone, etc.

Choosing your antenna correctly is almost as important as picking up the right radio. A good antenna will seriously improve your signal, so you need to give this choice some serious consideration.

We have already touched upon the topic of antennas in Chapter 3 but in order to make the best possible choice you will need to gather as much information as possible from all the available resources.

Although it may not seem as significant, make sure to pick quality feedline and connectors. Bad feedline can cost you a lot of your output, without you even knowing about it. A weak signal can cost you some contacts, and that's the last thing you want as a ham. Since you are building a station, do it right and make sure to cover all grounds.

If you decide to buy used cable, examine it carefully before the purchase. Check out for any dents, cracks or deposits of tar (from being outside). Make sure the inside of the cable is healthy, without any corrosion or discoloration. If there is a connector at the end of the cable, it may be difficult to access the inside, but don't be afraid to cut it off and install a new one after you buy the cable. A good cable is very important. Likewise, don't try to save money on connectors. You can find good ones at rather low prices.

*Jack Campbell*

There are other small things you will need to consider as well, but if you have reached this point in the book, then you are clearly very interested and motivated in becoming a ham operator. There is no lack of resources on hamming on the Internet, and as you become more involved with the community, you will no doubt find many experienced operators out there willing to lend you a hand.

Reading everything you can get your hands on will certainly help, but there is no real substitute for the hands-on experience that longtime hams have under their belts. Once you find the right conversations, don't be afraid to ask questions and show your interest. After all, everyone had to start somewhere, and people are interested in seeing the community grow.

Most of all, have fun with it. Hamming can be a really exciting and challenging hobby, which will help you get to know people from all over the world. Hams come from all walks of life, so you will, no doubt, meet some fascinating characters on your journey.

# Chapter 7

## Operating in HF Bands

The amateur radio was a vocation for experimenters and communicators at the beginning of the last century when it used lower frequencies than current HF (1.8 to 30 MHz). Today the tracks to amateur radio operators are situated beyond 250 GHz. Until 1970, the HF bands were the means used for communications coast to coast and international by amateur radio operators, commercial and government stations. These frequencies were so overloaded with commercial stations of all kinds in world traffic, as well as amateur communications city to city, dx, contests, utility networks and experiments.

More recently, this situation amateur was reversed by placing a repeater for the first time in the range of 2 meters, providing a new horizon in communications. Curiosity in intercontinental communication was also heightened by the ability of HF bands.

Next, we will make a short presentation of the characteristics of each range from 160 to 10 meters:

## 10 METERS:

A band, where new and old operators work in common, is 10 meters.

For ease in working with low power and relatively small antennas and their reach is global, this track is a favorite of beginners.

As a large spectrum band 28000-29700, many ways are used, CW, SSB, FM, satellite, digital modes, repeaters, among others. The propagation is usually open, even in times the low solar cycle. Many amateurs have managed feats in times of "lean spreads", working hundreds of countries during the few and short openings in the low solar flux times. 10 meters also shows us some characteristics of high bands VHF, not hear any signal in this band is not to say that there is no spread and the band is closed.

Short openings, but excellent, they are always possible. The most popular station 10 meters is a transponder with 100W and three small tri-band 10/15/20 or even a short monoband upon a roof. Linear amplifiers and high antennas are not really needed, however, always have your antenna as high as possible. Small rotors used in TV antennas will easily handle these sets of irradiating. By being a very broad operating band of 10 meters, it is recommended to use an antenna tuner for the case of working in the high and low band, ie making FM and CW. You can then very easily work a station from a friend at 130 miles or an island in the South Pacific. The absence almost always static, provide you operate weak signals or pile-ups and has a chance to be heard.

15 METERS:

Perhaps the most beloved of all the tracks, the conditions in this band are the most predictable for use in DX. The operation in fifteen meters is more common in low solar activity times. In these times, 15 meters behave better than the range of 10 meters or 12. The continental communications are possible during the whole day. This is one reason why most networks are concentrated in this range, including therein, emergency meteorological services, Red Cross, departments for foreign of USA and DX networks. In this band also the most common type of antenna is the three elements tri-band, but because of the special propagation conditions at certain times, you can work distant stations with relative ease, especially at sunrise and sunset.

20 METERS:

If any of the tracks can be labeled "war-horse", or "Track 4 Seasons", 20 meters certainly will be. In practice, this band brings together the best operators, the most powerful stations, and larger antennas. It is the preferred band of Honor Roll who use it for this purpose. Practically the ham radio world occupies this track, it being the most congested of all. It is the most popular for DX, SSTV, CW, digital operations and endless uses. Most operators use 100W and a tri-band antenna, managing to achieve success with incredible ease, so it is considered the elite band of ham radio. In times past, this range was attributed magical powers because they duped communication with all countries on the planet. Nowadays 20 meters remain the main means of communication especially DX in times of high solar activity or not. When propagation conditions are favorable, it is clear that the amount of heard stations may even frustrate us because of the difficulty of listening either. The solution then will come with a larger antenna, ex. six elements so will discriminate more the against unwanted signals. The communicated via long-path are also favored in this range.

40 METERS:

Here is the track that is more shared with other services. It is not uncommon to find commercial stations, underground and broadcast stations up to 500 KW operating in this location. A problem for amateur radio operators, but also an indicator of how it is spread. A broadcast station is a reliable indicator often. During the day, 40 meters is a mid-range band, up to 1250 miles, but the night we can contact any part of the world and you do not need 500 KW. 40 meters is the most important range for all types of competition, many hams have all countries of the world worked on this track. During the day, we have the rounds, at night, we have many DX nets trying to workstations from other countries operating in a split as an alternative to escape the broadcast and other interference.

80 METERS:

Things that many amateur radio operators think when you mention the bands of 80 and 160 meters, is always the size of the antennas, gigantic, and the number of noise found in those bands. But remember that the first amateur radio operators were restricted to operate in the range of 200 meters and below it! These tracks were very popular many years before our sophisticated equipment and computer programs to design antennas. Prior to popularize repeaters operated on 2 meters in the early 70s, a lot of local traffic, networks and emergency traffic was held in the range of 80 meters. For regional communications, narrow band antennas, shortened dipoles, vertical antennas and other arrangements combined with antenna couplers guarantee a regular performance. Dipoles full-size inverted-V provide good signal communicated to a few hundred miles, but the noise is also present. For long distance communications, DX, you may position your antenna horizontal dipole at least 15 meters (free space) or more. Everything is valid to get it intent, poles, trees, etc. To reach the mark of 250 or more countries at 80 meters, other arrangements are made: Antennas, slopers, vertical phase, devices and low-noise antennas for reception only etc ... Depending on your income, there are directional antennas 4 elements, for the paltry sum of

*Jack Campbell*

US $ 4,000.00 ... but they are not very popular. The 80 meters always represent an obstacle to those who are striving for the DXCC in 5 bands (5 Band DXCC) or diploma WAZ (Worked all Zones), not many amateur radio operators in the world can such a feat, which makes -If an unforgettable experience for those who did. What everyone has to say, it is that the latter zone worked, was an area of very little population and that there were very few hams there! To contact some areas and / or countries, it requires knowledge of propagation, gray - line, many hours of lost sleep, low noise, operating expeditions, good ears, especially PERSISTENCE! Once you have trained a lot in 80 meters you can try another challenge ... 160 meters.

160 METERS:

Known as the "Top Band", 160 meters is the range of experts. The amateur radio operators of this band are very dedicated to the study and experiment with all kinds of antennas.

Other operators, in addition, the experimenters who are in the range are the Dxs. During the day, 160 meters not offer us nothing but noise. You can find a typical strange noise of about 15 kHz which is the harmonic of a local TV device. During the summer, we can hear the static caused by storms hundreds of miles. During the night things change and many stations are heard, especially in spring and fall. We may hear strong signals, mainly in CW and also activity on the phone. There is a very wide variety of shortened antennas when installing with proper care, will provide good local contacts. Working Dx in 160 meters involves making many discoveries. It mainly involves hard work hunting. If you expect to find strong signals in the band any day of the year and at any time you turn on the radio, you're wrong! They are generally much weaker than those usually found in other low bands. The first step is the radical change to try to improve the signal / noise ratio in reception. As an initial solution installation Beverage antennas or frame.

One technique used is also the use of the 40 meters antenna to assist in copying weak signals. In general, we use an antenna for transmission and one for the reception.

The towers used to support the Hf antennas can be used to support the L reversed, another good option for this band. Keep in mind that the conditions can be different at 160 meters, such as QSB, noise and spread can be exactly the opposite on the same day and at the same time to those found at 80 meters.

## The Bands WARC 12/17 And 30 Meters

WARC, World Administrative Radio Conference in 1979, included the bands of 12, 17 and 30 meters and amateur radio operators were entitled to use in 1989. These new bands have the same characteristics in adjacent bands to their frequencies, but different propagation and the physical size of the antennas.

*Jack Campbell*

Another difference is the small portion allocated to communications. Each band has its advantages over time, for example, the band of 10 meters can be with the closed spread and 12 meters in perfect conditions at the same time. Although they are close to each other, the conditions may be quite the opposite. The 17 meters, a surprising track, for the conditions, are such as 20 meters, global reach and excellent propagation conditions most of the year. 30 meters, limited to using CW and digital modes and 200 W of power, provides good spreading conditions.

# Chapter 8

## DX - What Is It?

DX is the hobby of making two-way radio contact with distant stations in ham radio. Many DXers also attempt to receive written verification of receipt (sometimes referred to as "QSL" the ear stations). The Hobby's name comes from DX, telegraphic shorthand for the "distance" or "distant".

**DX station**

Although the classic definition of DX is "away" today usually means contacting amateur radio stations in distant places. In the HF bands (also known as short-wave), DX stations are those in foreign countries. In the VHF / UHF bands, DX stations may be within the same country or continent, since it is a VHF contact long distance (radio to radio), without the help of a satellite repeaters or other devices.

For the award, other areas of policy that countries can only be classified as "DX countries." For example, the

French protectorate of Reunion Island in the Indian Ocean is counted as a DX country, even if it is a territory of France. The rules for determining what is a DX country can be quite complex and to avoid possible confusion, amateur radio operators often use the term entity rather than country. In addition to entities, some awards are based on groups of islands in the world's oceans. In the VHF / UHF bands, many amateur radio operators seek awards based locators Maidenhead Grid (GRID LOCATOR).

For the more rare locations, DX-peditions are often organized to allow amateur radio operators "work a new location."

## Communication DX

DX communication is communication over long distances using the ionosphere to refract the radio transmission beam. The beam back to the Earth's surface, and can then be reflected back to the ionosphere into a second jump. Ionospheric refraction is often feasible only for frequencies below about 50 MHz, and is highly dependent on weather conditions, time of day, and the cycle of sunspots eleven. It is also affected by solar storms and some other solar events that can change the Earth's ionosphere, ejecting a shower charged particles.

The angle of refraction places a minimum on the distance at which the beam will be refracted first return to Earth. This distance increases with frequency. As a result, any station using DX will be

# Ham Radio

surrounded by an annular dead zone where they can not hear other stations or be heard by them.

This is the phenomenon that allows reception of shortwave radio to occur beyond the line of sight limitations. It is used by amateur radio operators, the shortwave broadcasters (as the BBC and Voice of America) and others. This is what lets you listen to AM stations (MW) from locations away from its location. It is one of the backups to miscommunication long-distance satellite when its operation is affected by electromagnetic storms from the sun.

## Operating Rules DX

Observe the rules below, possibly you will have a nice long DX-man career:

1. Call the DX station that you want to talk ONLY after that it call CQ, QRZ, transmit SK or VA or the equivalent in phone.
2. Never call a station DX:
    2.1 The station QRG where he is working, until you are sure that the QSO in progress was completed.
    2.2 When he transmits KN, AR, CL or equivalent in phone.
    2.3 After he makes a directional CQ unless you are in the region or area requested. Ex .: CQ CQ Africa, the US is not in Africa, so do not answer, wait for him to call North America (NA).
3. Keep within the limits of frequencies allowed by our legislation. Some DX stations broadcast out of the frequencies of the permitted.

86

4. Pay attention to the instructions of the DX station, 10U or 10 UP means call 10 Khz above his QRG; 15D or 15 DOWN, mean 15 Khz down, etc.
5. Be honest in their reporting because many DX stations will depend on your report for station settings and equipment.
6. Keep your always clean signal. handling clicks, variations of QRG, power use above the permitted modulation excess (SPLATTER) will bring you only bad reputation.
7. Listen and call the DX station that interests you. Calling CQ DX is not the best way to hunt calls.
8. When several stations waiting to work a "call", avoid asking for it "MEET A FRIEND". Let your friend plays with others.
9. If possible before entering pile-up, while you study the best way to hunt the "call" look properly hear the call sign, your QSL Manager (or be who answers the QSL) etc... to avoid loss of time and / or possible errors.

Finally, the good impression you leave on our foreign friends is worth more than you think.

## Choosing your QRG

Some little things about propagation are necessary you know: four are the main factors that determine the propagation characteristics:

1. Frequency in which you are operating

2. Time of day or night

3. Season (summer, winter, fall and spring).

4. Cycle of sunspots.

Selecting the appropriate band relies heavily on three other factors. For example, 80m (3.5 to 3.8 Mhz) at noon, summer, in the best condition of the sunspot cycle is certainly the worst possible choice as the same band, at midnight, in winter, in the worst condition sunspots, produces excellent DXs. Quickly you learn by experience and observation when you are operating in a given band in the best DX.

Be aware that in many bands # 2, 6 and 10m are Beacons (signaling stations) that are in operation to indicate the openings of said bands.

Note between 28.2 and 28.3 MHz (10 meters) and around 50.110 MHz (6 meters). commercial stations near the ends of the amateur bands also provide a reliable reference of propagation openings, just remember that most of these stations operate with far superior powers to allowed for us and, as a result, they can be heard well before opening the check to the extent necessary to allow communications between hams.

## What don't To Do

It is illegal to broadcast music. This can happen when we have our house full of people.

It is normal to have music in the house and also other people listening to the radio and / or TV, K7 recorders or playing musical instruments.

Music is no problem when operating CW, but can easily be transmitted when operating telephony (SSB,

FM, AM, etc.). Stories and ribald jokes are also banned from amateur bands.

Releases for financial business are also not allowed in order to sell or buy equipment is not good practice.

When in doubt, don't do anything before to determine whether or not is legal!

# Chapter 9

# Failures Common In Amateur Radio

- The radio does not turn on: Check the feed connection was done correctly. If everything is normal to check if not broke a fuse of the radio, or if there is no bare wire. In other cases, consult a specialist (deteriorated transistor).
- Burning fuse so that puts the appliance: bad contact inside or outside (connection) of the device, check the wires and poles. In a mobile installation, do not try to replace the fuse with a higher intensity.
- Reception absence: The "squelch" may be all open, or "RF" potentiometer in minimum. Check also if the device is not in a position to "public address", and if the microphone is on. Indeed, CB devices cannot receive signals if the microphone is not mounted.
- Intermittent reception: Check the feed connection, and also the coaxial cable jacks to the PL antenna and transceiver. Also, check the earthing quality.
- Signal saturation at the reception: If the captured station is very close, this phenomenon is normal. Adjusting the "RF gain" is if you have this control, or ask the other

party to decrease your "mic-gain". If these provisions do not have any effect, you can treat yourself to a fault in the speaker.
- Reception and weak noise, even when the volume is at maximum: Readjust welding of coaxial to the socket device PL, and verify that this outlet is tight to the base of the antenna.
- Reception several signal channels from another station: Using a linear amplifier by a station located in the neighborhood and affecting adjacent channels. Find the frequency of issue of this disturbing and ask him to eliminate the problem.
- Receiving a station, but unable to answer you: Adjust the antenna. It can, however, this is a too remote station, using an antenna better and less suffering from obstacles.
- Parasites strong and perhaps regular at the reception: proximity to industrial or medical devices with resonance in the 27 MHz, high-voltage lines or of radio beacons.
- Absence when you press the microphone button: bad contact the microphone at its outlet or its cable.
- Bad modulation: "Mic Gain" too deeply, antenna regular or damaged microphone to fail. In fixed installations, feed inadequately.
- Transmission without modulation: If you can get the problem lies probably in own microphone or its cable.
- Variable modulation: coaxial cable badly welded to the outlet PL, whose power connection is poorly executed.
- Inability to contact stations, unless very close: poorly regulated antenna or simply bad propagation conditions. In the latter case, there is nothing to do.
- In mobile installations, noise when the engine works: Connect the filter "NB - ANL" or anti parasite the radio.
- The appliance heats exaggeratedly: rates too high standing waves. Remedy the situation quickly.

- Inability to get a reduced rate of standing waves: bad grounding. Adjustable antenna rod must be cut (with a maximum precaution) with optimal dimensions for a good line.

*Jack Campbell*

# Chapter 10

## Operational Technical

Operational technical practices are known by amateur radio operators, so if in doubt as to the expressions, see:

1. Before making a CQ, make sure that the frequency (QRG) is unoccupied.

2. When you challenge a CQ, tune your equipment "beat zero" in the QRG colleague in order to facilitate their reception. The only exception to this rule is in the case of operation "split" - when the station transmits at a frequency and receive on another - previously announced. Also, keep in mind that our tracks are, increasingly, becoming small on the growing number of ham radio.

3. Identify at least every 5 minutes and at the beginning and end of a QSO. These are internationally accepted rules.

4. The station, in order, it is to use the frequency is the one that must meet the other to call and identify a space between cambium. The reason for this is to keep the sequence of a round.

*Jack Campbell*

5. Never try to pass "over" another station. First, because it is illegal! And second, because it hurts everyone.

6. If you suspect that you are modulating with another station, turn off the PTT or VOX and listen to make sure.

7. It is a signal of poor operational practice leaving the frequency "to the rightful owner," because it usually creates some doubt or confusion soon after.

8. If the frequency is transferred to a station to make a quick call to someone, the conversation between them should be as short as possible or both stations should change QRG.

9. The word "break" is reserved strictly for emergency traffic.

10. Do not operate on frequencies not you are allowed.

11. Keep up-to-date with the amateur radio law. Always keep in mind the conditions under which got the privilege of being an amateur radio operator.

12. Do not use the tracks for advertising commercial activity, political or religious.

13. Refrain from acts that are characterized as a commodification of amateur radio, or trade for amateur radio operators. Besides being illegal, their behavior is being observed by mates.

14. Each ham has the right to seek to achieve the objectives legally covered by their license. However, it has the duty to avoid causing inconvenience or interference to others.

15. If there is a narrow band segment that is used for international communications (DX), avoid using it for chats.

16. Respect the segments of the bands for the various operational practices: CW, digital, satellite, etc., because there is enough space for a harmonious and peaceful coexistence of all ham radio modalities.

17. In local daily chats, give preference to the use of low bands (40 and 80 meters) and then use the high band VHF / UHF (50, 144 and 430 MHz).

18. Usually, communicated long distance are preferred over sites.

19. Eager to get a contact (QSO), avoid unduly trample the QRG, occupying it before completion of the previously established contact. If you notice that a colleague began the challenge to a CQ, expect the contestation. As the work of the fellow who called CQ, if you have not heard the other station, it is up to you to make your call and try to contact.

20. In DX and "pile-up", respect the natural order of QSO, avoiding benefit this or that station. In exceptional cases, this may be admitted only if the favored season is for low power (QRP).

*Jack Campbell*

21. CW use the signs internationally recommended, especially at the end of each exchange; in order to prevent impatient tapping could harm the QSO.

22. It is considered that a statement is true when the two stations have changed the target and reports correctly.

23. When a station is a directed CQ, adding the geographical area with which to speak (CQ Asia, Europe CQ, CQ Africa, CQ North America), only one should dispute it when you are in the region or country indicated by whom He made the call. Otherwise, it will only slow, it is not the goal of the caller.

24. Be brief, precise and concise in DX contacts. In the "pile-up" then give the area code, report and nothing else.

25. CW never pass up the speed with which it was contested.

26. Do not make endless CQ. Make short calls. Most CW operators do QSY (change frequency) to hear general called (CQ) endless.

27. DX, established contact, quote indicative one or at most twice.

28. DX, repeat only words and data "key". Do not pass on QSZ (repeat all the words).

29. telegraphy respects the spaces, not amend the letters. The rhythm is more important than speed. Remember that our tracks are for amateurs.

30. Do not bother to pass quickly. Use moderate pace, though, as perfect as possible. A telegraph operator is also judged by its ability to receive and not only for its speed and cadence transmission.

31. The CW high-speed operation can and should be used since both stations are able to do it and understand perfectly.

32. When you hear a mate CW issue the letters "CL" at the end of QSO, do not insist. Is it a lack of courtesy to the other station, as the "CL" means that the station intends to terminate the operation (QRT).

33. Communicated "pile-up" avoid asking for information, as the DX station always passes the data of the respective Manager from time to time, as well as specifications on your station.

34. Never make interrogation when contact with a special code. The only appropriate question is "MANAGER PSE? (Please, manager)" or "QSL INFO? - Information about who or where to send the QSL (card).

35. Listen well before trying to get a call. Faced with a "pile-up", avoid offering indicative own without knowing who it is and then ask your callsign: "PSE UR CALL? '.

*Jack Campbell*

36. If the DX station operates in "SPLIT" and you can not afford to do it, forget the call, but will be disturbing others with their call, without the slightest possibility of holding the contact.

37. The ends of each track are used to difficult communications, DX. Always try to remember that. So do not use the beginning of each sub-band for contests, QRS contacts or local chat.

38. Before accessing a repeater, first, listen to familiarize yourself with the features of its operation.

39. To initiate contact, you may communicate that you are in attendance, stating your callsign.

40. Take a break between transmissions, ie, a space between the exchanges. This will allow other hams can ask the opportunity to speak or communicate its presence in frequency.

41. You may avoid actuating several times a repeater without identifying themselves, only to hear the sound of the connection.

42. Make short exchanges ensure the durability of the repeater equipment and the "space" needed to ask other hams or opportunity to pass messages needed. The monolog can prevent someone with an emergency, use the repeater or the QRG. If the monolog is long enough, he can "burst" the timer repeater, "knocking her down."

43. Use statement simplex wherever possible. If you can complete a QSO in a direct frequency, there is no need to keep busy repeater and prevent others from using.

44. Please use the minimum power required to maintain communication. In addition to not force the equipment, minimizes the possibility to actuate another more distant repeater that we may use the same frequency.

45. Some repeaters are equipped with "autopatch" (connection to the telephone network) that, properly used, it provides many facilities. So you do not abuse the privilege to don't lose.

46. Amateur radio Operators should behave with full respect for the laws, especially those governing the Amateur Radio Service.

47. Our obligations, before the other fellow radio amateurs, are not limited to regulatory provisions. The most important thing is to use common sense and comity, to share frequencies with which we are destined.

*Jack Campbell*

# Chapter 11

# Several Expressions Conventional

There are many abbreviations used by amateur radio operators in radiotelegraph communications, especially when in contact with other countries. Some of them, most commonly used are listed below and are generally derived from words in the English language.

**RST Code**

| Number | R – Readability |
|---|---|
| 1 | Unreadable |
| 2 | Barely Readable |
| 3 | Readable with difficulty |
| 4 | Readable without difficulty |
| 5 | Perfectly readable |

| Number | S Strenght |
|---|---|
| 1 | Faint signs perceivable |
| 2 | Very Weak signals |
| 3 | Weak signals |
| 4 | Fair signs |
| 5 | Signs very reasonable |
| 6 | Good signs |
| 7 | Signs moderately strong |
|  | Strong |

*Jack Campbell*

|   | Very strong |
|---|---|

| Number | Tone |
|---|---|
| 1 | 60 Hz or less |
| 2 | Very strong AC tone |
| 3 | Strong AC tone, rectified but unfiltered |
| 4 | Rough note, some hint filtering |
| 5 | Filtered rectified Ac, strong ripple modulation |
| 6 | Filtered tone, defined traces of ripples |
| 7 | Near pure tone, ripple or modulation of traces |
| 8 | Tom almost perfect, slight modulation features |
| 9 | Perfect tone, without ripple traits or modulation of any kind |

## CW Abbreviation

| ABBREVIATION | MEANING | ABBREVIATION | MEANI |
|---|---|---|---|
| AA | All After | OB | Old Boy |
| AB | All Before | OC | Old Cha |
| ABT | About | OM | Old Ma |
| ADEE | Addressee | OP | Operato |
| ADR | Address | OPR | Operato |
| AA | All After | OT | Old Tim |
| AB | All Before | PBL | Preamb |
| ABT | About | PKG | Packag |
| ADEE | Addressee | PSE | Please |
| ADR | Address | PT | Point |
| AGN | Again | PWR | Power |
| AM | Amplitude Modulation | PX | Press |
| ANT | Antenna | R | Receive |
| BCI | Broadcast | RC | Ragche |

103

## Ham Radio

|      | Interference          |      |                          |
|------|-----------------------|------|--------------------------|
| BCL  | Broadcast listener    | RCD  | Receive                  |
| BCNU | Be seeing you         | RCVR | Receive                  |
| BK   | Break in              | REF  | Refer to                 |
| BN   | Between, Been         | RFI  | Radio F Interfer         |
| BT   | Separation            | RIG  | Station Equipm           |
| BTR  | Better                | RPT  | Repeat                   |
| Bug  | Semi automatic key    | RTTY | Radiote                  |
| C    | Yes, Correct          | RST  | Readab Strengt           |
| CFM  | Confirm, I confirm    | RX   | Receive                  |
| CK   | Check                 | SASE | Self add stampe envelop  |
| CKT  | Circuit               | SED  | Said                     |
| CL   | Closing Station, Call | SEZ  | Says                     |
| CLBK | Callbook              | SGD  | Signed                   |
| CLD  | Called                | SIG  | Signatu                  |
| CLG  | Calling               | SINE | Persona or nickn         |
| CNT  | Cant                  | SKED | Schedu                   |
| CONDX| Conditions            | SRI  | Sorry                    |
| CQ   | Calling any station   | SS   | Sweeps                   |
| CU   | See you               | SSB  | Single S                 |
| CUL  | See you later         | STN  | Station                  |
| CUM  | Come                  | SUM  | Some                     |
| CW   | Continuous Wave       | SVC  | Service                  |
| DA   | day                   | T    | Zero                     |
| DE   | From, From this       | TFC  | Traffic                  |
| DIFF | Difference            | TMW  | Tomorr                   |
| DLD & DLVD | Delivered       | TKS & TNX | Thanks              |
| DN   | Down                  | TR & TX | Transm                |
| DR   | Delivered             | T/R  | Transm                   |
| DX   | Distance              | TRIX | Tricks                   |

104

| | | | |
|---|---|---|---|
| EL | Element | TT | That |
| ES | And | TTS | That is |
| FB | Fine business | TU | Thank y |
| FER | For | TVI | Televis interfer |
| FM | Frequency Modulation, From | TX | Transm Transm |
| GA | Go ahead, Good afternoon | TXT | text |
| GB | Goodbye, God Bless | U | You |
| GD | Good | UR | You're Y |
| GE | Good Evening | URS | Yours |
| GESS | Guess | VFB | Very Fi Busines |
| GG | Going | VFO | Variabl Freque Oscillat |
| GM | Good Morning | VY | Very |
| GN | Good Night | W | Watts |
| GND | Ground | WA | Word A |
| GUD | Good | WD | Word |
| GV | Give | WDS | Words |
| HH | Error sending | WKD | Worked |
| HI HI | Laughter | WKG | Workin |
| HR | Hear | WPM | Words minute |
| HV | Have | WRD | Word |
| HW | How, Copy? | WX | Weathe |
| IMI | Repeat, say again | TXVR | Transce |
| LNG | long | XMTR | Transm |
| LTR | Later | XTL | Crystal |
| LVG | Leaving | XYL, YF | Wife |
| MA & MILLS | Milliamperes | YL | Young |
| MSG | Message | YR | Year |
| N | No, Nine | 73 | Best Re |
| NCS | Net Control Station | OB | Old Boy |

*Ham Radio*

| ND | Nothing Doing | OC | Old Cha |
|---|---|---|---|
| NM | No More | OM | Old Ma |
| NR | Number | OP | Operato |
| NW | Now, Resume transmission | OPR | Operato |
| LVG | Leaving | OT | Old Tim |
| MA & MILLS | Milliamperes | PBL | Preamb |
| MSG | Message | PKG | Packag |
| N | No, Nine | PSE | Please |
| NCS | Net Control Station | PT | Point |
| ND | Nothing Doing | PWR | Power |
| NM | No More | PX | Press |
| NR | Number | R | Receive |
| NW | Now, Resume transmission | RC | Ragche |

*Jack Campbell*

# Conclusion

Thank you again for downloading this book!

Hopefully, these pages have helped you learn everything you need to know to become a ham yourself. While there are some topics that could be explained in more detail, this book should give you clear direction in which to look if you want to become a serious amateur radio operator.

The intention is not to make you an expert, but give you a good base, a range of initial information so you can have a good start in the wonderful world of ham radio and especially of Radiotelegraphy. This is just the beginning because there are many variables in Ham Radio, a very large number of options that it would be impossible to cover them all.

What comes next is entirely up to you. You can look for more information, online and offline, or you can simply take the leap, using the information provided here and learn as you go.

Have fun and enjoy the experience of meeting new people using the technology that is often neglected in this modern period. There is a whole new world hidden in the invisible radio waves just waiting to be discovered. And, as you can see, becoming a part of it

*Jack Campbell*

is not that hard at all. We really hope this book will be useful to you!

Finally, if you enjoyed this book, then I'd like to ask you for a favor, would you be kind enough to leave a review for this book on Amazon? It'd be greatly appreciated!

*Ham Radio*

*Jack Campbell*

*Ham Radio*

Printed in Great Britain
by Amazon